Learning About the Movement of the Sun and Other Stars with Graphic Organizers

Isaac Nadeau

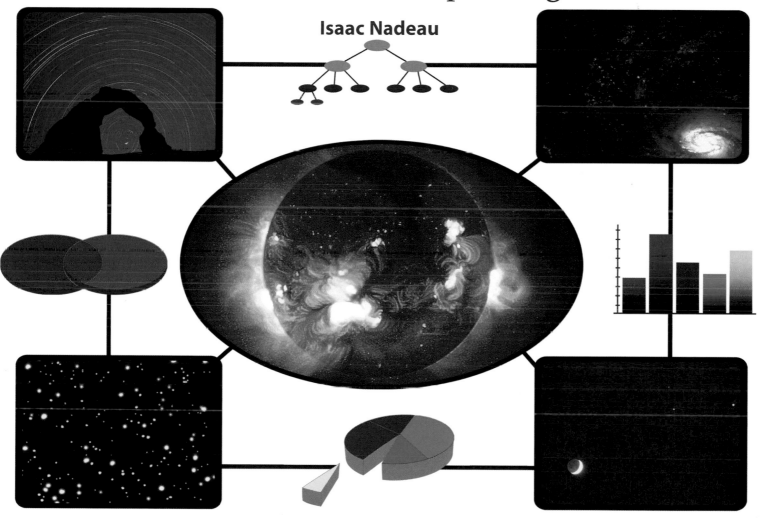

Rosen Classroom Books & Materials

New York

Dedicated to Gramps and Grandaddy

Published in 2006 by The Rosen Publishing Group, Inc.
29 East 21st Street, New York, NY 10010

First Edition

Editor: Natashya Wilson
Book Design: Mike Donnellan

Photo Credits: Cover (center), p. 7 (top left) © DigitalVision; cover (top left) © Paul A. Souders/CORBIS; cover (top right) © Royalty Free/CORBIS; cover (bottom left) © NOAO/AURA/NSF; cover (bottom right) © Roger Ressmeyer/CORBIS; p. 7 (bottom right) © Gary Braasch/CORBIS; p. 12 illustration by Michael DeGuzman.

Nadeau, Isaac
        Learning about the movement of the sun and other stars with graphic organizers / Isaac Nadeau.
                p. cm. — (Graphic organizers in science)
        Includes index.
        Summary: This book uses texts and graphs to explain the movement of the sun and other stars.
        ISBN 1-4042-2805-5 (lib.)—ISBN 1-4042-5040-9 (pbk.)—ISBN 1-4042-5041-7 (6-pack)
        1. Sun—Juvenile literature   2. Stars—Juvenile literature   [1. Sun   2. Stars] I. Title   II. Series
        QB521.5.N33 2005   2003-021540
        523.7—dc21

Manufactured in the United States of America

Contents

The View from Here   5

What Is a Star?   6

The Sun   9

Polaris and the Southern Cross   10

Shapes in the Sky   13

Stars and Planets   14

Stars Moving in the Universe   17

Shooting Stars   18

Bright Travelers   21

Astronomy, Then and Now   22

Glossary   23

Index   24

Web Sites   24

Cause-and-Effect Chart: Earth's Movements

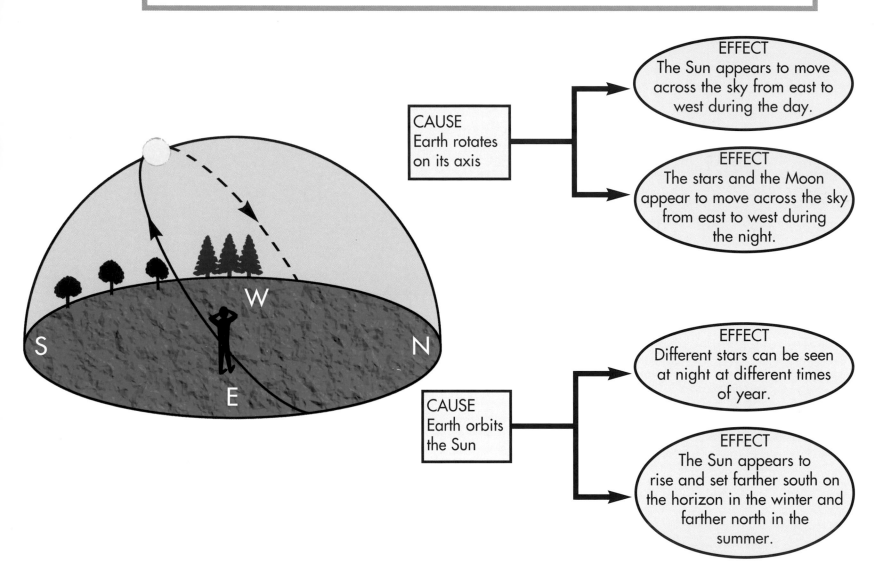

CAUSE
Earth rotates
on its axis

EFFECT
The Sun appears to move across the sky from east to west during the day.

EFFECT
The stars and the Moon appear to move across the sky from east to west during the night.

CAUSE
Earth orbits
the Sun

EFFECT
Different stars can be seen at night at different times of year.

EFFECT
The Sun appears to rise and set farther south on the horizon in the winter and farther north in the summer.

The View from Here

The Sun, the Moon, and the stars seem to move across the sky each day and night. The Sun rises in the east, moves in an **arc** across the sky during the day, and sets in the west. The Moon and the stars move in similar arcs across the sky at night. Earth seems to stay still while the sky moves in a circle around it. However, these apparent movements of the Sun, the Moon, and the stars are really caused by the **rotation** of Earth.

It takes Earth about 24 hours, or 1 day, to complete one rotation on its **axis**. From Earth, this rotation makes the Sun, the Moon, and the stars appear to move through the sky.

Graphic organizers are written tools that can help you to study. In this book, graphic organizers are used to help you learn about the movement of the Sun and other objects in the sky.

Right: *This cause-and-effect chart shows that Earth's movement causes the Sun and other stars to appear to move. The items on the left side are causes. They make the effects on the right side happen.* Left: *This diagram shows what the movement of the Sun looks like to a person on Earth. As Earth turns, the Sun appears to move from east to west.*

What Is a Star?

Stars are giant balls of hot gas. The main gas in stars is called **hydrogen**. As the gas burns, it gives off heat and light. This light allows us to see the stars from Earth. If Earth did not move, the stars would seem to stay in one place. However, **astronomers** have found that each star is moving on its own path through the endless space we call the **universe**. There are **billions** of stars. They look tiny from Earth, but that is only because they are so far away. Most stars are huge. The Sun is an average-size star. It has a **diameter** of about 865,000 miles (1.4 million km), which is more than 100 times the diameter of Earth. The largest stars, called supergiants, may be 300 times as large as the Sun. Stars give off different colors of light, depending on how hot they are. The colors of most stars can be seen only through a **telescope**.

This graphic organizer is called a concept web. To make a concept web, write a subject in the middle of a piece of paper. Then write down everything you know about that subject around it. As you learn more facts, add them to the web. This concept web is about stars.

Concept Web: Stars

THE SUN
The Sun is a star. It is the star that is closest to Earth. Other stars can only be seen at night because the Sun's light makes it too bright to see them during the day.

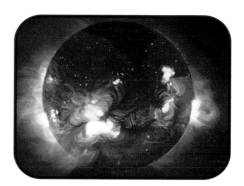

COLORS
Stars give off different colors. The hottest stars are blue-white. The coolest stars are red-orange. Medium-hot stars are yellow. The Sun is a yellow star.

STARS
Stars are made of hot gas. The gas burns, making stars shine.

MOVEMENT
Stars are moving through space. We cannot see the movement because they are so far away from Earth. On Earth, stars appear to move across the sky in arcs. Their apparent movement is caused by Earth's rotation.

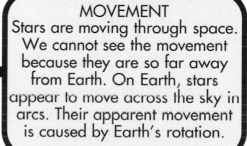

SIZE
Stars are huge. The Sun is a medium-size star. It is 100 times bigger than Earth. The biggest stars, called supergiants, may be 300 times the size of the Sun.

KWL Chart: The Sun

What I Know	What I Want to Know	What I've Learned
• The Sun seems to move in an arc across the sky every day.	• Why does the Sun seem to move when it is really Earth that is moving?	• When you stand still on Earth, you are still moving at the same speed that Earth is spinning. Earth is moving you past the Sun, but, to you, the Sun appears to move.
• Sunlight feels warm.	• How hot is the Sun?	• The surface of the Sun, called the photosphere, is about 10,000°F (5,500°C). The corona is a ring of fiery gases that burst outward from the photosphere. The heat of the corona is about 1.8 million°F (1 million°C). The heat in the center, or core, of the Sun can reach 27 million°F (15 million°C).
• The Sun rises and sets at different times every day.	• Why do days get shorter and longer?	• The angle of Earth's axis causes days to grow longer and shorter as Earth moves around the Sun. Because of the angle, the part of Earth that gets more daylight changes from the northern half to the southern half over the course of one year. Days get longer in one half of Earth as they get shorter in the other half.

The Sun

At about 93 million miles (150 million km) away, the Sun is the closest star to Earth. If you could fly to the Sun at 100 miles per hour (161 km/h), it would take you more than 100 years to get there. It takes heat and light from the Sun just 8 minutes to reach Earth. The heat and light provide Earth with **energy**.

As Earth rotates on its axis, it also moves in an **orbit** around the Sun. It takes Earth about 365 days, or 1 year, to orbit the Sun. Over the year, the Sun's path across the sky appears to change. For example, in the Northern **Hemisphere**, the sunset is slightly southwest on the **horizon** in the winter and slightly northwest in the summer. In the winter, the Sun rises later and sets earlier than it does in the summer. These seasonal changes happen because Earth rotates at an angle to the Sun.

This is a KWL chart. To use a KWL chart, choose a subject for your chart. Begin by filling in the left column with what you know. Then ask yourself what you want to know about that subject. Write your questions in the middle column. As you study the subject, write down what you learn in the third column. This KWL chart shows facts about the Sun.

Polaris and the Southern Cross

For centuries, astronomers and explorers in the northern half of Earth have used a star called Polaris, or the North Star, to guide them. If you could sit at the North Pole and watch Polaris for an entire night, Polaris would appear to move in a very small circle straight above you while all of the other stars circled around it. This is because Polaris is located almost directly above the North Pole. People find Polaris to figure out which way is north. When you face north, west is to your left. East is to your right, and south is behind you.

Polaris cannot be seen from the southern half of Earth. People there can find their way using a group of four stars called the Southern Cross. Together the four stars look like a kite. The tip of the Southern Cross points toward the South Pole.

A sequence chart is a graphic organizer that can help you to learn the order of the steps in a process that has a beginning and an end. This sequence chart teaches you how to find Polaris. Begin at the top. Then follow the arrows from step to step. Remember, you must be in the Northern Hemisphere to find Polaris!

Sequence Chart: Finding Polaris

Find the group of stars called the Big Dipper. It looks like a square pot with a bent handle.

↓

Find the two stars that make the front edge of the pot. They are in the bottom and top right corners.

↓

Draw an imaginary line straight through both stars from the bottom star to the top star.

↓

Continue your imaginary line above the Big Dipper until you reach a bright star. That is the North Star, also called Polaris. When you are facing Polaris, you are facing north.

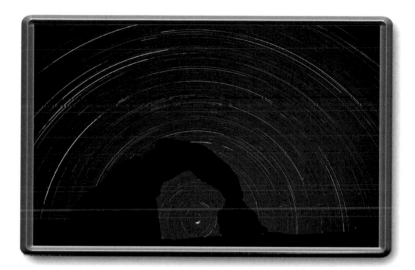

This picture was taken by holding the lens of a camera open for a long time. It shows the night sky as seen from Arches National Park in Utah. The bright dot in the middle is Polaris. The streaks around it are the apparent paths of other stars.

11

Map: Constellations

Shapes in the Sky

A **constellation** is a group of stars that people think make a shape in the sky. Since ancient times people have named constellations and created stories about them. For example, the ancient Greeks said that the constellation Orion is a hunter who is being chased across the sky by another constellation, Scorpio. Scorpio is part of a group of constellations called the Zodiac. The Zodiac includes 12 constellations that are lined up with Earth's orbit. A new Zodiac constellation comes into sight every month. Some constellations, such as Orion and the Zodiac, can be seen from both hemispheres. Cassiopeia and the Seven Sisters are two constellations that can be seen only from the Northern Hemisphere. The Southern Cross and the Austral Triangle are two that can be seen only from the Southern Hemisphere.

Maps show the locations of places and things. This is a map of the night sky. These are constellations you can see in the Northern Hemisphere from spring to summer. The Big Dipper is in the northwest corner. A line leads from its pot to Polaris. Some of the Zodiac constellations can be seen in a line from southeast to southwest.

Stars and Planets

When you look at the night sky, every bright dot above you may look like a star. However, a few of the dots may be planets in our **solar system**. You can tell which is which by watching for the movement of planets in the night sky. Stars never change position with other stars. Planets move slowly to different positions among the stars night after night. There are nine planets, including Earth, in orbit around the Sun. Five of them, Mercury, Venus, Mars, Jupiter, and Saturn, can be seen from Earth without a telescope. Planets also look different from stars. Most stars twinkle. Planets do not. The easiest way to tell a planet from a star is to see whether or not it twinkles. Unlike stars, planets do not give off their own light. They **reflect** starlight. Also, though you cannot see it, planets have a hard center. Stars do not.

Top: This picture shows Venus, Mars, Jupiter, and the Moon. Venus is the brightest dot. Mars is the dimmest. Bottom: A Venn diagram shows how two things are different and how they are alike. The features the two things have in common go in the middle, where the circles overlap. This Venn diagram compares stars and planets.

Venn Diagram: Stars and Planets

Stars | Planets

- Twinkle

- Do not change positions with other stars

- Do not have a solid center

- Give off light and heat

- Look like dots of light in the night sky

- Appear to move in arcs from east to west during the night

- Do not twinkle

- Slowly change position among the stars night after night

- Have a solid center

- Reflect starlight

Compare/Contrast Chart: Movement of the Sun and Other Stars		
	Apparent Movement	Real Movement
The Sun	From Earth, the Sun appears to move across the sky in an arc from east to west during the day. In the United States, it appears to rise farther south on the horizon in winter and farther north on the horizon in summer.	The Sun is turning on its axis and moving in an orbit around the Milky Way galaxy. It is not moving around Earth. Earth is moving around the Sun.
Other Stars	From Earth, the stars appear to move in arcs from east to west during the night. They appear to stay in fixed positions in relation to other stars.	Each star is turning on its axis, as are Earth and the Sun. They are orbiting their galaxies. They are also moving away from one another. They are so far away from Earth that we cannot tell they are moving.

Stars Moving in the Universe

Earth, the Sun, and all the stars and the planets are part of an unending area of space called the universe. Most scientists believe that the universe formed about 14 billion years ago. At that time, everything in the universe was packed into a tiny space. This **dense** ball burst in an explosion called the big bang. Ever since the big bang, everything in the universe has been moving outward in all directions. Stars also move in two other ways. They rotate on their axes. It takes the Sun about 27 days to make one turn on its axis. Stars also move in orbits. Most stars are part of a **galaxy**. Every galaxy has a center of **gravity** that pulls stars and other things in orbit around it, just as the Sun pulls Earth and other planets around it. Our solar system is in a galaxy called the Milky Way. There are about 200 billion stars in the Milky Way.

Top: *This is the Milky Way galaxy. The Sun moves through the Milky Way at about 150 miles per second (241 km/s). It takes the Sun about 225 million years to orbit the Milky Way. Bottom: This graphic organizer is called a compare/contrast chart. It compares the apparent and real movements of the Sun to those of other stars.*

Shooting Stars

Have you ever looked up at the night sky and seen bright lights streak across it? We call these streaks shooting stars, but they are not stars at all. They are meteors. They begin as bits of rock and dust that have been floating in space. These bits are drawn toward Earth by gravity. As they hit Earth's **atmosphere**, they become so hot that they burn up, causing a streak of light. As do planets, meteors orbit the Sun. Their orbits cross the path of Earth's orbit once in a while. Earth passes through several large groups of these bits of rock and dust each year. During these times, meteor showers occur. Hundreds of meteors can be seen in one night. Astronomers can **predict** when some meteor showers will take place, so you can find out the best time to watch. A meteor shower called the Perseids occurs on August 11 or 12 each year.

This graphic organizer is called a timeline. Making a timeline can help you to learn when events happened. It will also teach you the order in which they happened. This timeline shows the order of historical events that involve the Sun, stars, and meteors. The earliest event is at the top. The most recent event is at the bottom.

Timeline: The Sun, Stars, and Meteors

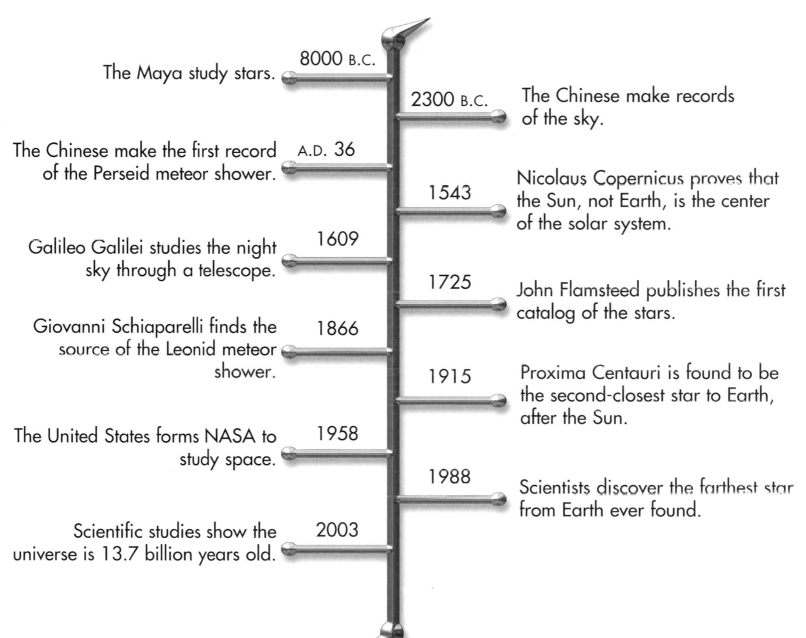

The Maya study stars. — **8000 B.C.**

2300 B.C. — The Chinese make records of the sky.

The Chinese make the first record of the Perseid meteor shower. — **A.D. 36**

1543 — Nicolaus Copernicus proves that the Sun, not Earth, is the center of the solar system.

Galileo Galilei studies the night sky through a telescope. — **1609**

1725 — John Flamsteed publishes the first catalog of the stars.

Giovanni Schiaparelli finds the source of the Leonid meteor shower. — **1866**

1915 — Proxima Centauri is found to be the second-closest star to Earth, after the Sun.

The United States forms NASA to study space. — **1958**

1988 — Scientists discover the farthest star from Earth ever found.

Scientific studies show the universe is 13.7 billion years old. — **2003**

Chart: Objects in the Night Sky

Object	Features	Movement	Best Time to See
Comets	Look like bright balls of light with long streaks of light trailing behind them.	Orbit the Sun. They pass close to the Sun, then travel far away before coming back.	Check on the Internet for the best dates to look for a comet.
Meteors	Look like streaks of light shooting across the sky.	Orbit the Sun.	At night.
The Moon	Appears to change shape over about 28 days, going from a dark new moon to a full moon and then back.	Turns on its axis. Orbits Earth. It takes the Moon about 28 days to orbit Earth.	Easiest to see at night. Sometimes the Moon can be seen during the day as well.
Planets	Look like bright dots of light that do not twinkle.	Move very slowly in their orbits around the Sun.	Mercury and Venus are best seen just after sunset or just before sunrise. Mars, Jupiter, and Saturn can be seen all night at different times of year.
Satellites	Look like bright dots moving quickly across the sky. They do not twinkle.	Orbit Earth.	At night. Satellites are easiest to see during a new moon.
Stars	Look like tiny, twinkling dots of light in the night sky.	Move outward. Turn on their axes. Orbit their galaxies.	At night. Stars are easiest to see during a new moon.

Bright Travelers

On clear nights, you may see as many as 2,500 stars. You can also see many other things moving through space, such as **satellites**, comets, and the Moon. Satellites look like tiny dots of light moving quickly across the night sky. They are machines, made by people, that are in orbit around Earth. Some satellites take pictures from space. Satellites are also used to track the weather and to send telephone and television signals.

A comet is a chunk of dust and snow in orbit around the Sun. Most of the time, comets are at the far edge of the solar system. Every so often, a comet's orbit will take it closer to the Sun. At this time, the comet can be seen from Earth. Comets have long tails of dust and gas that stretch out behind them. From Earth they look like bright balls with long tails of light trailing behind them.

A chart is a graphic organizer that is used to organize all kinds of facts about related subjects. The subjects are listed in the far left column. Facts about each subject appear in the row next to it. The type of fact is listed at the top of each column. This chart shows facts about many different objects you may see in the night sky.

Astronomy, Then and Now

Ever since people first looked up at those bright dots of light that decorate the night sky, they have had questions about the universe and its movements. Ancient astronomers believed that Earth was at the center of the solar system and that everything orbited it. In the 1500s, Nicolaus Copernicus said that the Sun was at the center of the solar system and that the Earth and other planets orbited it. In the 1600s, an Italian astronomer named Galileo Galilei was the first person to study space using a telescope. He discovered that moons were in orbit around the planet Jupiter. In the twentieth century, Cecilia Payne-Gaposchkin found that stars such as the Sun are made mostly of hydrogen and helium gas. Today scientists are sending satellites to the edge of the solar system and beyond to continue the search for knowledge about the universe.

Glossary

arc (ARK) A curved path.

astronomers (uh-STRAH-nuh-merz) People who study the Sun, the Moon, the planets, and the stars.

atmosphere (AT-muh-sfeer) The layer of gases around an object in space. On Earth, this layer is air.

axis (AK-sis) A straight line on which an object turns or seems to turn.

billions (BIL-yunz) Thousands of millions. One billion is 1,000 millions.

constellation (kon-stuh-LAY-shun) A group of stars.

dense (DENTS) Closely packed together or thick.

diameter (dy-A-meh-ter) The measurement across the center of a round object.

energy (EH-nur-jee) The power to work or act.

galaxy (GA-lik-see) A large group of stars and the planets that circle them.

graphic organizers (GRA-fik OR-guh-ny-zerz) Charts, graphs, and pictures that sort facts and ideas and make them clear.

gravity (GRA-vih-tee) The force of attraction between matter.

hemisphere (HEH-muh-sfeer) One half of Earth or another sphere.

horizon (huh-RY-zun) A line where the sky seems to meet the earth.

hydrogen (HY-droh-jen) A colorless gas that burns easily and weighs less than any other known matter.

orbit (OR-bit) A circular path.

predict (prih-DIKT) To make a guess based on facts or knowledge.

reflect (rih-FLEKT) To throw back light, heat, or sound.

rotation (roh-TAY-shun) The spinning motion of a planet around its axis.

satellites (SA-til-yts) Machines in space that circle Earth.

solar system (SOH-ler SIS-tem) A group of planets that circles a star.

telescope (TEH-leh-skohp) An instrument used to make distant objects appear closer and larger.

universe (YOO-nih-vers) All of space.

Index

A
Austral Triangle, 13
axis(es), 5, 9, 17

C
Cassiopeia, 13
comets, 21
constellation, 13
Copernicus, Nicolaus, 22

G
galaxy, 17
Galilei, Galileo, 22
gravity, 17–18

H
helium, 22
hydrogen, 6, 22

M
meteors, 18
Milky Way, 17
Moon, 5, 21

N
North Pole, 10
North Star, 10

O
Orion, 13

P
Payne-Gaposchkin,
    Cecilia, 22
Perseids, the, 18
Polaris, 10

S
Scorpio, 13
Seven Sisters, 13
Southern Cross, 10, 13
South Pole, 10

Z
Zodiac, 13

Web Sites

Due to the changing nature of Internet links, PowerKids Press has developed an online list of Web sites related to the subject of this book. This site is updated regularly. Please use this link to access the list: www.powerkidslinks.com/gosci/sunstar/